cloverleaf books™

Planet Protectors

Go Easy on Energy

Lisa Bullard

illustrated by Wes Thomas

M MILLBROOK PRESS · MINNEAPOLIS

For Vicki and Steve Palmquist —L.B.
For Becky —W.T.

Millbrook Press
A division of Lerner Publishing Group, Inc.
241 First Avenue North
Minneapolis, MN 55401 U.S.A.

Website address: www.lernerbooks.com

Main body text set in Slappy Inline 18/28. Typeface provided by T26.

Library of Congress Cataloging-in-Publication Data

Bullard, Lisa.
        Go easy on energy / by Lisa Bullard ; illustrated by Wes Thomas.
            p.   cm. — (Cloverleaf books. Planet protectors)
        Includes bibliographical references and index.
        ISBN 978-0-7613-6107-7 (lib. bdg. : alk. paper)
        1. Energy conservation—Juvenile literature. I. Thomas, Wes, 1972– ill.
    II. Title.
    TJ163.35.B85 2012
    333.791'6—dc22   2010053302

Manufactured in the United States of America
1 – BP – 7/15/11

TABLE OF CONTENTS

I'm Tyler. Mom calls me
Sleepyhead Tyler.
I'd love to stay curled up with
my dog, Pete, every morning.

But I hurry to catch the bus. Then Mom doesn't have to drive me.

Driving Mom's car takes energy.

6

We're learning about energy in school. People use it to heat buildings and light rooms. People even used energy to make my lunch box.

Chapter Two
Energy Problems

Using all that energy causes problems. Mom's car runs on energy that comes from **oil**. Cars that use oil energy make the **air dirty**.

I want to use less oil energy.

That will help save the Earth.

9

Electricity is another kind of energy. Dad told me our electricity comes from the **power plant.** It's a big building across town. The power plant burns a rock called coal.

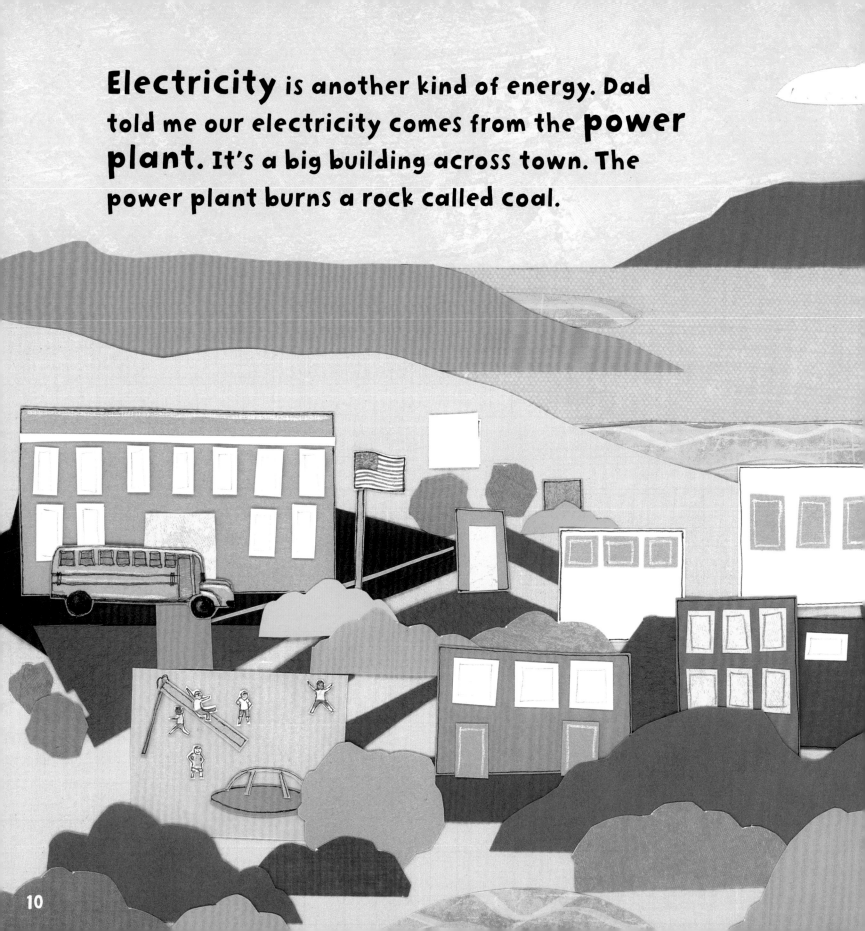

Burning **coal** makes the air dirty too. **YUCK!**

Coal

Plants and animals that died long ago were buried underground. Dirt and rocks piled on top. The plants and animal remains were pressed down. After millions of years, they changed. Some turned into coal or oil.

In school, we watch a movie. It shows us ways to get clean energy from water, sunlight, or wind. But these energies can cost a lot. And we can't get them everywhere.

Wherever our **energy** comes from,
we should try to use **less of it.**

Wind turbines gather wind energy. Wind farms have many of these machines. Wind energy is very clean energy. It doesn't make the air dirty. But some people think turbines are ugly. Some people worry that they hurt birds. And not all places get a lot of wind.

13

Chapter Three

Saving Energy

At recess, I suddenly remember. I was on Power Patrol today. But I forgot to turn off our classroom lights. **Uh-oh!**

So I decide to be Power Patrol at home.

I turn off the TV to save energy.

See all the things with power cords? They use a lot of electricity.

Turning something off does not always stop it from using energy. Unplugging the power cord is important too. But remember to check with a grown-up first! Some things should not be unplugged. You also need to be careful with plugs. Always make sure you turn something off before you unplug it. Ask a grown-up to show you other plug safety tips.

Walking instead of driving helps the Earth too. Walking uses **people power.** Or dog power! They're both clean power.

18

But walking makes me hungry. Did you know people get their energy from food? I'm glad we're having my favorite energy for dinner—**pizza energy!**

19

Maybe tonight, I'll dream up more ways to **use less energy.**

GRRRRRR

Do you think you can help **save energy** too?

Then I can find another way to
save the planet tomorrow.

Play the Power Patrol Game

Here's a picture of part of my house. See me and Pete? I'm eating pancake energy. Pete's eating dog-food energy. Many of the other things at home also need energy. I'm still Power Patrol at my house. I made it into a game. I made a list of all the energy users I could find. It's really long!

You can play this game too. How many things do you see that need energy to work? When you're done, turn this book upside down. You can check my list to see if you missed anything. Then look around your own house to play the game.

Here's a hint: Batteries store energy. So look for things with batteries too.

fan, blender, coffeemaker, microwave, CD player, mp3 player, radio, light, telephone, refrigerator, and clock

battery: a container filled with chemicals that produces energy

coal: a black or dark brown rock that comes from plants buried underground millions of years ago

electricity: a form of energy. Electricity is used to power many things, such as lights and TVs.

energy: the ability to do work, such as moving, playing, and growing

power cord: a cord that electricity can move through

power plant: a place that makes electric energy

wind turbine: a machine that makes electricity when wind spins its blades

BOOKS

Hock, Peggy. *Our Earth: Clean Energy.* New York: Scholastic, 2009. Learn more about clean energy from things such as water, sunlight, and wind.

Bradley, Kimberly Brubaker. *Energy Makes Things Happen.* New York: HarperCollins, 2003. This book explores different kinds of energy and fuel, sources of energy, and how energy moves from one thing to another.

WEBSITES

Alliance to Save Energy: Energy Hog
http://www.energyhog.org/childrens.htm
Learn more about energy use through an online game. This game will teach you how you can stop being an "energy hog."

California Energy Commission: Energy Quest
http://energyquest.ca.gov/index.html
This website has movies, games, and more information to help you save energy.

U.S. Department of Energy: Kids Saving Energy
http://www.eere.energy.gov/kids/
Visit this site for more ideas of ways you can save energy, as well as games and riddles.